The Total Safety Committee CHECKLIST

A Step by Step Handbook for Safety Committee Success

MATTHEW A. FORCK

The Total Safety Committee Checklist
A Step by Step Handbook for Safety Committee Success

Printed in the United States of America

Published by:

K-Crof Industries LLC, *dedicated to building SAFE people through*
Kcrof *writing, speaking and giving!*

In 2002, Bill Dampf, a safety professional for the utility company with which I was a first line supervisor, called me. He asked me if I wanted to serve a temporary role on his safety staff. I'm not sure he really knew what he was getting when I enthusiastically responded, 'Yes!' That was the proverbial 'crack in the door' that started the ball rolling. Five years later I earned my Certified Safety Professional (CSP) designation, spoke in Baltimore on safety leadership and Chicago on safety awareness. It all started with Bill, a guy who took a chance on me and for that I will be eternally thankful. And while it is only a small token of thanks, I dedicate this book in his honor… Thanks Bill!

Table of Contents

The Total Safety Committee Checklist, an Introduction

Recently I purchased a new power washer. I rationalized the expense because I was going to use it to clean my deck. I could buy the washer and stain and do the work myself or hire someone to do it for the same cost. For a 'tool guy' it wasn't a tough choice, all I needed to know was 'power washer!'

As with most tools and children's toys these days, the power washer came in a box with those famous words, "Some Assembly Required." I was fearful of directions with small font written by an attorney to protect against liability. What I found however was much different. On a large single piece of paper, in full color, I found step-by-step assembly instructions. Not only was it printed on a nice big piece of paper but it also had big numbers and easy to read instructions. In less than 15-minutes I was washing the deck!

After a weekend of power washing the deck, the car, the siding, the dog, (just kidding about the dog), I returned to work. My role is motivating people to make safe choices and to empower leaders and teams that make safety go, safety committees. After reflecting on these easy-to-follow assembly instructions, I believe that we can do better with instructions for safety committees, too. There are a ton of instructional materials out there for committees. There are step-by-step instructions, how to guides and safety committee manuals for dummies. But these instructions can be difficult to follow. All too often they are scattered on internet sites, and in policy books. They can be buried in procedures; PowerPoint slides, training documents and all too often detailed with small print. While this document may not be able to include each and everyone of these finer points, it is a single, inclusive book with a simple purpose; to set out a simple, results driven and easy to read set of step-by-step instructions that can guide a safety committee to success for an entire year!

Note: as with products we purchase, they often include instructions on models other than what you have. There will be references used in this document that may not apply to your model (Safety Committee). And, if it doesn't apply then simply skip over to the next item.

Best of luck with the Assembly of your committee!

Getting the Right Tools out of the Tool Box:

Before we can go to the checklist and mark off safety committees 'to dos' we first must establish that we have the right tools needed for a successful assembly. In safety that means we have the right alignment, process, systems and a common language for safety committees success. Specifically, there are four primary tools needed. They are as follows;

- Tips for On Target Teams; an outline for a comprehensive management system, which can support a team for long-term success.
- The Safety Committee MAP Process (MAP means Monthly Awareness Plan), which puts measureable systems in place to address the primary hazards faced in any work environment.
- ISMAs, Involved Safety Meeting Activities, a means to re-invent safety meetings and increase retention three-fold in the process.
- The Untapped Secret to Selling Safety, how a team can bring safety awareness to a whole new level through effective yet seldom used sales techniques.

It is these processes, definitions and systems that will be referenced throughout the monthly checklists.

Tools Needed Before Assembly Begins:

TIPS FOR ON-TARGET TEAMS — THE FIVE KEYS LEADERS INSTILL INTO EFFECTIVE SAFETY TEAMS

Which quote best describes your safety team? Does your team often, "keep minutes but lose hours?" Milton Berle. "A committee is composed of people who individually can do nothing, who come together to conclude that nothing can be done," by an unknown author. "A committee is ten people doing the work of one," unknown. Or finally, would you select 'D', all of the above?

Chances are that you have a safety team. Actually, chances are you have many safety teams within your organization. And, chances are those teams are greatly under performing. The Behavior Based Safety revolution of the last couple of decades introduced us to the Safety Committee, a team of individuals gathering at regular intervals to do something. The problem is that the team and management alike are not always sure what the team should be doing or how it should get done. In truth, some of these teams do very well. Some teams enjoy brief moments of achievement surrounded by months of mediocrity. Most, however, underachieve, month-in-and-month-out to the frustration of managers.

To be fair to our teams however, it's not all their fault. If most organizations are completely honest, our best safety teams happen by chance and are the exception and not the rule. For our under-achieving teams, we have failed to give them the key tools that would enable their success; until now. Employ these five tips for on target teams and revolutionize your safety team's performance.

TIP I-A SHARED BELIEF SYSTEM, SBS

Description: It probably wouldn't surprise you to learn that there are over 60,000 table-saw injuries a year. In fact, in the time it takes to read this page there will be more fingers lost to a table-saw. Most teams, when faced with these facts, would conclude that injuries are 'just part of using the saw.' A response to this conclusion may be an aggressive team goal to reduce injuries on table-saws by 25%! Does that sound familiar?

A few years ago, however, a team met to discuss table-saw injuries. One team member believed that these injuries could be prevented-and he was able to sell this belief to the entire team. "Understand, what we believe," says Max DePree, former CEO of Herman Miller Co., Inc. "precedes policy, procedure and practice." Armed with a shared belief system, this team produced different results than any previous team. They found that they could place a small electric charge on the blade along with a meter. The meter continuously measured voltage on the blade and any voltage drop, such as when a hand touched it. When the meter detected a drop, it trigged a breaking mechanism that stopped the blade 10-times faster than a vehicle air bag is activated. The result was no harm to the operator and the elimination of injuries. This was only possible because the team believed in the same highly challenging yet achievable end.

Trap: There is an elephant in the room that no one is talking about, and that is the team's 'shared belief system.' The truth is that a shared belief system is the foundation of a team's success. We can have a talented team, provide outstanding management support, follow through on the other four tips for on-target teams, hang high glossy colored safety posters, etc. but if we fail to engage, understand and challenge the shared belief system, our team will not produce positive change. The trap is failing to discuss beliefs and agreeing on a shared system, something like, 'all table-saw injuries are preventable.'

Tip: Address the elephant in the room by holding honest discussions on individual beliefs regarding safety and injury prevention then establish a team belief statement. This is an over-arching statement about what the team believes is possible in safety. Beginning with the words, "We believe..." can be the powerful starter your team needs for success.

TIP II-CRYSTAL CLEAR DIRECTION (CDC)

Description- "If you don't know where you are going," Yogi Berra says, "You'll end up somewhere else!" We would never hail a cab and then tell the cab driver, "I'm not sure where I need to go, just go." Instead, when we get in a cab, we have a direction, purpose and an expected arrival time;

after all, the meter is running. Unfortunately, too often we stick our safety teams in a cab and tell the driver 'just drive' because we fail to give them crystal clear direction. The best way to provide our teams with direction along the ride is through a specific end statement (Hackman). This is a statement of team destination but not the route (means) in which to get there. Management's role is to give the team this CDC, the team's job is to use the provided tools, training, knowledge and talent to get there.

Trap: Management's job is to offer a specific end statement. The team's job is to outline their course of action based upon that end statement. The trap is that management, not wanting to interfere with the team, fails to examine the team's action plan and challenge or suggest additional action items. Just as a team should demand a specific end statement, an action plan to meet that end should be required of the team.

Tip: Generally safety teams have a chairperson and secretary and the remaining team members just show up and eat donuts. Once the specific end statement is communicated, have the team identify the roles needed for the team's success. Then, each team member is assigned one or more of these specific roles. This will engage each team member to a higher level. No longer can they just show up; with a specific role, they will have to show up and be ready to report and participate.

TIP III-CLEAR STRUCTURE

Description: In early December of 2006, Mid-Missouri was blanketed by nearly 24-inches of snow. In an area of the country where a half-foot of the white stuff can close schools, 2-feet immobilized the region. In addition, many warehouses and factories were damaged when roofs collapsed under the weight of the snow. One of the more intriguing roof cave-ins however was a large horse barn. While most roofs failed in the first 24-hours, this one failed five days later. After an investigation by insurance officials, it was determined that the roof was built to withstand the 24-inches of snow, when that snow was distributed evenly. There were, however, several periods of melting and re-freezing and the snow shifted, sliding and gathering along the mid-point on each side, causing failure.

"Every organization," Stephen Covey says, "is uniquely designed to exactly produce the results it achieves." Nothing is more accurate when referring to our teams. Clear structure can be defined as "what is acceptable on this team." It's the written or unwritten set of norms and rules that allow the team to function well or cave-in under its own weight.

Trap: When teams are initiated, teamwork and assignments (weight) are more or less distributed equally among the team; everyone is eager and willing to step in and do their part. Over time, if structure and rules are not clear, some team members will disengage and fail to complete assignments, roles and responsibilities. In the end, one or two team members will be carrying the weight of the entire team, leading to a collapse.

Tip: Establish rules and norms and find fun ways to enforce them. For example, one team I served on had a rule that anyone late for a team meeting or late returning from a break was fined $5.00-all of the money going to charity. It was a fun way to establish the team value of promptness…by the way, I was only fined once and was never late again after that!

TIP IV-TEAMS ARE WELL SUPPORTED

Description: "An army runs on its stomach," is a quote used frequently and has a deeper meaning than just food. It refers to all of the supports such as food, fuel, clothing, medical, training, communications, transportation, logistics, etc. that are needed to make an army mission successful. In a similar way, our teams 'run on their stomachs.' There are a host of supports that need to be in place for safety teams to flourish and succeed. I think the five key safety team supports are as follows; time, budget, training/skill development, clerical/office assistance and feedback/recognition.

Trap: Blaming poor team performance on 'the team.' At the end of the day, these are our teams and a team that is not performing up to par may be more of a reflection in what we have put in there 'stomach' than on the team members themselves. The trap is that we may 'write off' a team because of their poor performance instead of asking the tough question, what supports did I fail to give this team?

Tip: Ask the team what they need! I realize it's a novel idea, but too often we assume the team has what they need and never ask. Ask the team leaders key questions such as, 'what keeps getting in your way?' or 'what is slowing you down?' Let them lead you to a better supporting role.

TIP V-SUCCESSFUL TEAMS ARE WELL COACHED

Description: It should not be a secret in a safety world dominated by the behavior based safety model that coaching or feedback is one of the five keys to a successful team.   There seems to be a lot of pressure on safety staff and management leaders to be great coaches. Well, forget about the pressure to be great and simply strive to be a 'GOOD' coach. GOOD means, Get in the game-you can't be effective if you are not in touch with the pulse of the team. Offer feedback-too often we observe team inadequacies, but fail to offer effective feedback to get the team back on course. Be Optimistic-since the team will often reflect the energy of the coach, we must be and remain up beat. Finally, be Determined to make a difference-understand your role as a coach is to change the team when sometimes the team is striving to 'stay the same.'

Trap: I've seen it time and time again, the trap is a simple and blatant failure to coach. Too often, we assume that since the team consists of a cross section of our leaders that they don't need our feedback…that assumption leads to ineffective teams. In truth, safety teams generally consist of field or floor employees. They are subject matter experts but not astute on the inner workings of successful teams.

Tip-To ensure consistent and effective coaching, design a feedback sheet. The feedback sheet can have questions which evaluate the team's specific end statement action plan, structure, support and coaching. Have safety staff or a management leader fill the sheet out during each meeting. After the meeting, have the team leaders discuss the evaluation tool along with specific adjustments that need to be made to make the team more effective for the next meeting.

In the end, John Maxwell said it best, "It takes team work to make the dream work." Teams are a part of our business and safety culture. Making them an effective part of this process is the next step to our goals of a zero-injury work place. Are your teams on target?

Have a Process-The Safety Committee MAP

THE SAFETY COMMITTEE MAP (MONTHLY ACTION PLAN) — YOUR COURSE TO SAFETY COMMITTEE SUCCESS

To create real change within an organization, one must alter resources, values or procedures, this according to Clayton M. Christensen author of the New York Times Bestseller, *An Innovator's Solution*. To be successful, safety committees need the time and budget (resources). The organization must establish a value base that fosters safety. But, once those are in place, we still need a means (procedure) to cultivate safety committee success.

FINDING THE POT HOLES...

In the past, safety committees have done as well as can be expected. We have asked them to set goals in an attempt to ensure their proactive success and we have encouraged and supported these goals. But in failing to offer a specific action plan we have asked our committees to 'play ball' without defining the field. Imagine a baseball team attempting to 'play' without a defined field, a pitchers mound, home plate, batters box, bases, fowl lines or assigned positions. The team can play hard but without a defined field it is hard to win.

This is similar to what our committees have been doing. Instead of being consistently proactive and awareness driven, they become reactive, engaged in few if any safety awareness activities. Most exhaust a great deal of time discussing items that are not safety related. Few have a plan for the next month. Many, in an attempt to do something, grab issues that should have already been addressed by management. And, management lets the committee have the issue, thankful that they are doing something. They are playing hard, they are just not sure of the 'field.'

A wise truth reads, "The arrow that hits the bull's eye is the result of a hundred misses." Our committees have shot many arrows, some hitting but most missing. Over the last few years, I have taken keen note of what arrows have hit and which ones missed. In so doing, I introduce the Safety Committee MAP (Monthly Awareness Plan), a strategy that will hit the bull's-eye month after month.

BEGIN THE DRIVE...

To begin the MAP process, brainstorm using the following questions. Where are our workers taking shortcuts? What are we doing that gets us hurt? What safety rules are neglected? Which safety rules are the most important? What does our previous peer observation data tell us? What training do we need this year? What needs does your community have? What holidays do you celebrate?

Twenty minutes of brainstorming can produce a hundred or more possible themes. Pick your favorite twelve. These will be your safety themes for the year (one per month). You and the committee have the discretion to change a theme if needed, but at least you have established your framework for the year.

SHIFTING INTO GEAR...

For each theme the committee should develop a MAP, each MAP should include the following.

Two Safety Activities. A safety activity is something done outside of a safety meeting that brings awareness to safety and your monthly theme. It might be a note taped to a locker, stickers or awareness items in trucks or at workstations etc.

An Involved Safety Meeting Activity; which is a safety meeting that involves the entire group. It can be a game, activity or exercise that gets everyone involved. There are references that can help with this such as my book, *ISMA –101 Ways to Get Your People Involved.*

An outside safety speaker. This is an individual from outside of the work group that can share his/her insights on the monthly theme. It might be a subject matter expert from the community (police officer, doctor) or an individual from another part of the company.

Distribute at least one type of safety educational material during the month; this is an item such as a quiz, cross word puzzle or other educational game. It can be used during a safety meeting to educate the group on the monthly theme. These can be easily found on the Internet or created using inexpensive 'word game' software.

Finally, hand out at least one safety trinket. A safety trinket is a tangible item that is given to employees to bring continued awareness to the monthly safety theme. For example, a Crunch Candy Bar can be used to remind co-workers not to be caught in a 'line of fire' crunch.

Once each activity, speaker, involved safety meeting etc. is chosen assign responsibility for that activity to members of the committee. This ensures the task is completed on time.

WHY UNFOLD THE MAP?
The MAP concept does seven unbelievable things for your organization.

The MAP shows caring. The soft side of safety is about relationships. Our people don't care what we know until they know we care. People know that you care when you are there for them, out front and visible. The MAP framework fosters caring and relationships.

Involvement. So many committees have asked me in the past, how do I get my co-workers involved? We typically think someone is involved when they come forward to deliver a safety meeting or write a safety quiz. Let's face it, that doesn't happen often. When following the MAP, everyone is involved. When I hand an employee a safety trinket or include him/her in an involved safety meeting activity, he/she just became involved in safety. And, involvement equals participation, which leads to ownership and that equals results!

The MAP fosters safety committee teamwork. Did you know that a Belgium Draft Horse could pull 8,000 pounds by himself? If you team two horses together they can pull 18,000 pounds. If you train them to work as a team, they can pull 25,000 pounds. Too often, committees are lead by one or two individuals. The MAP engages the entire committee, in the end, making them more effective.

The MAP keeps focus. Committees will be focused on their monthly activities and are more likely to divert concerns that can be handled outside of the committee to the proper parties. You can't hit a target you can't see, but with this plan you've defined your targets and your arrows, it's a win/win.

The MAP is high safety energy awareness. 'Reaction' and 'creation' have the same eight letters. When you react you respond in the same fashion in which you have always responded. Yet when you create, you think before you act. Maintaining high-energy safety awareness makes our workers think, thus creating safe situations instead of reacting to old habits.

The MAP communicates to the work group. When the committee is being reactive, the work group often doesn't even know they exist; the committee is contributing nothing to their work group. When following the MAP framework there will be no doubt the committee is out in front and that workers are actively supporting the safety of their peers.

The MAP changes behaviors in a positive direction. Since the monthly themes are picked based on a high number of incidents in a certain area, past injuries or weak safety habits, these monthly activities continue to encourage safe behavior, challenging and changing at-risk behavior.

DRIVING HOME SAFE...

Have you ever wondered why the white stripe is painted on either edge of a narrow two-lane highway? Well, when oncoming headlights, fog, snow or rain blind a driver, all he needs to do is find and focus on the white line. If the driver doesn't have a focus, he can easily loose control and drive off the road. If he focuses on the yellow centerline, he can easily cross that line into on coming traffic. Although visibility is poor and the driver can't see ahead, he knows that continued long-term focus on the white line will get him to his destination safely. The Safety Committee MAP (Monthly Action Plan) is our 'saving' white stripe. Open the MAP and drive knowing that you are changing your safety culture, one MAP at a time.

Safety Committees are a tremendous investment in time and money for organizations yet most of the time, the return on the investment falls short. Safety Committees fail to function at a high level because they simply don't know how. Following the Safety Committee MAP (Monthly Action Plan) offers a simple easy-to-use plan to ensure Safety Committee Success, month after month.

Re-Inventing Safety Meetings... this time for Results!

ISMAS (INVOLVED SAFETY MEETING ACTIVITIES)

"When you can do the common things of life in an uncommon way you will command the attention of the world."
-George Washington Carver

At nineteen years old, Isaiah Thomas left Indiana University after just two years and was the fourth pick in the 1982 NBA draft. Much to his dismay, the lowly Detroit Pistons selected him. Detroit was a team without a tradition or identity, in fact they had won just 16 of 82 games in the previous season. Isaiah was not used to losing. Just months earlier, he lead his team to the NCAA national championship. In public, Isaiah tried to put his best face forward, but in private he had serious reservations about joining such a poor performer.

Instead of being a victim or 'just doing time' with the Pistons organization, he began studying great teams, paying particular attention to what it took to build a great team. In his recent book entitled, *The Fundamentals*, he said this on building a great team; "Long term success of sports dynasties comes when an organization is dedicated to bringing out the best in all of its people. You don't create lasting bonds with rah rah speeches and slogans plastered on the locker room walls. You build a team by getting everyone involved from the equipment manager to the coaching staff and owner. You help every member of the organization understand how he or she contributes to the pursuit of long-term mutually beneficial goals. You establish a shared vision and a team culture with a standard of excellence and achievement. I learned that when you set the bar high and let everyone know that they are expected to push their talents to the limit at every practice and every game, your people rise to that standard of expectation."

While we're not playing basketball, building a dynasty is building a dynasty whether it is sports, business or a safety culture. We have to be dedicated to bringing out the best in our people and getting everyone involved.

THE MYTH OF INVOLVEMENT:

Most leaders, managers and safety committees realize what Isaiah noted above, success is about involvement. After all, involvement equals participation, which equals commitment, which equals ownership. Ownership equals results. Yet most are confused and frustrated about how to get co-workers involved. The traditional mode of involvement goes like this. The safety leader will stand in front of his co-workers during a safety meeting and ask, beg or bribe someone to lead future meetings. When absolutely no one volunteers our leader feels defeated.

We need to turn our ideas about involvement in a safety meeting format 180 degrees. If we plan our safety meeting around an activity in which everyone participates then we have just involved the whole group. We have done so without waiting for a volunteer and without the group even realizing what has happened. This involvement in many respects is better than the traditional for two reasons; you have just involved everyone (not just a single meeting leader) and the activity is a better teacher then the traditional sit and listen safety meetings.

I HEAR IT AND FORGET, I SEE IT AND I REMEMBER, I DO IT AND I UNDERSTAND.

Traditional safety meetings are filled with videos and reading material, all 'hear' or 'see' activities. Information is soon forgotten if received at all. Building a successful 'dynasty' comes when "an organization is dedicated to bringing out the best in all of its people" and building safety success is no different.

Bringing out the best in our people means working with your people on a level of 'understanding.' That means in safety meetings we must strive to 'do' instead of simply hear or see. "Doing" an involved activity will lead to understanding and over time, success.

THE MYTH OF QUANTITY VERSUS QUALITY...

There was a man once, who after years of neglect, overeating, lack of exercise and basic disregard to his body, decided that he should change. He went to his doctor who performed an extensive and exhaustive physical. Two days later this man sat down with his doctor. The doctor said, "I have some good news, you're not overweight, you are exactly six inches too short." And, "you seem to be in great health for someone who is 82 years old." They both laughed knowing the man was only 45 years old. Finally, the doctor said, "all kidding aside, you weigh 300 lbs, your ideal weight is 200 lbs. I suggest exercise, eating right and reducing stress."

Returning home the man set a goal to lose 100 lbs over the course of the next two years. He was so confident that he could loose 4.16 lbs a month; he didn't even begin dieting for the first 27 days. Two years later the man weighted 200 pounds. He was exercising and eating right and enjoying life.

When this man weighed 300 pounds, he had a lot of quantity. He lost 33% of his total body mass leaving only quality. Most safety meetings are bloated from years of neglect, lack of planning and lack of a clear goal. Each week they seem to drag on for forty minutes or even an hour, there is much quantity but not a lot of quality. In shifting your safety meetings to ISMAs, you will be focused on a clear goal and teaching through doing. In short, you will be moving from quantity to quality.

MOTIVATIONAL VERSUS OPERATIONAL
Do you need operational or motivational ISMAs? You need both, and this is why. Safety can be divided into three parts, body, mind and soul, let me explain.

Body safety encompasses the physical tasks we must perform. For example, in the utility business, an electric line worker has a physical tasks of running a chain saw, climbing a pole, operating a backhoe, installing a wood cross arm etc.

Mind safety is the knowledge we possess that allows us to do our job. It includes the safe work rules, safe work practices, standards etc.

In general, we don't take risks (shortcuts) because of a lack in body skills or knowledge. A shortcut by definition is a failure to realize ones true worth, then subsequently acting in a manner that places one at-risk to an injury.

In short, we need both operational and motivational ISMAs. Operational activities will re-train on safety rules and safe work practices. Motivational activities are aimed at increasing worker motivation, understanding of values, self worth and sense of family so that they will not place themselves or coworkers at risk.

THE NUTS AND BOLTS...

For operational ISMAs, break down a safety rule asking these questions, how can we do it, solve it or ask questions about it? Using an example from the utility industry, one rule reads, "Before setting a meter, the worker shall ensure, proper voltage, no back-feed or grounds in the meter base." Traditionally, the safety meeting facilitator would review this by reading the requirement and then verbally reviewing the proper tests to be performed to ensure compliance, "a real sleeper." But, if we plug in the question, "Can we do it?" we realize there is a better way. Breaking the group into smaller groups of four, we can bring in a meter base for each group and have them perform the proper checks while wearing all of the proper PPE. Instead of reading to them, each person has the opportunity to review it 'hands on;' talk about understanding!

For motivational ISMAs, check out my book ISMA –101 Ways to Get Your People Involved or try an Internet search using key words 'games trainer's play.'

IT'S HARD SIMPLY BECAUSE IT'S NEW!

Isaiah Thomas discovered making a winning team took hard work and determination, but over time he did it, winning two NBA championships with the Detroit Pistons. His secrets of bringing out the best in people and getting everyone involved are the same key concepts at the heart of each ISMA. The ISMA concept isn't hard, just different. Hard work, determination and the willingness to make this small change will change your culture forever...let's play!

 Selling Safety, as Never Before!

THE UNTAPPED SECRET TO SELLING SAFETY

You may be thinking, "I'm not in sales, I'm a safety professional!" Well, you are half right, you are a safety professional. The truth is that if we are in safety, we are in sales too. The fact of the matter is that we are at the mercy of our ability to sell, no matter how "tight" the presentation, our education or the facts surrounding an issue. Everyday we are still in a position where we have to make the sale in order for a positive change to take place. And, the better we are at selling, the greater our results. So, how do we learn the secret of selling safety? Well, from the master of selling himself, Joe Girard.

Joe, born into tremendous poverty on November 1, 1928, was raised in a Detroit ghetto. He grew up a hard worker and was dedicated to get ahead. In his twenties and married, he began to make his mark in the Detroit real estate market. He was doing well until the bottom dropped out and Joe fell with it, broke and unemployed.

Determined not to let his young family down, he relentlessly searched for a job. Finally, he talked a sales manager of a Chevrolet dealer into giving him an opportunity. On his first day, he was told to sit in the background and only help if everyone was busy. Later that day Joe stayed to lock up. As luck would have it, a woman entered the showroom just as the doors were being locked. Two hours later, Joe sold her a car…and he was about to make history.

Joe did well selling cars, very well. He was successful using a number of sales techniques but maybe the most unorthodox was the concept of handing an item to a prospective buyer. He explains in his book, *How to Sell Anything to Anybody,* that he had desk drawers full of items. Everyone who entered his office received a small token. He had a drawer of stuff for women, one for men and a third for children. When people entered his office, the equation was unbalanced. In the customer's mind, Joe was going to be taking from them through a sale. In order to balance the equation, put the customer at ease, build trust and a relationship, Joe would give each customer an item from his desk drawer, each customer, each time, no exceptions.

Selling Safety Awareness…Willie Hammer, in his instructive and educational book, *Occupational Safety Management and Engineering*, Fourth Edition, writes, "Intermittent safety efforts are generally ineffective. It is necessary to maintain an almost continual program of keeping personnel alert to safety practices." He continues by citing examples of awareness opportunities, "Small folders or booklets available from safety organizations, insurance companies, and the federal government can be given to each employee as he or she enters the plant or at other appropriate times. Slips with printed safety messages can be added to pay envelopes or attached to pay checks. Place mats or napkins with interesting messages on safety and accident prevention." The point is that after certain safety fundamentals are in place such as a formal safety process, management safety accountability, inspections, etc., awareness is key.

So, how do we sell safety awareness to our work force? Part of the sale is within all of the traditional approaches such as safety meetings, posters, slogans, bulletin board postings, competitions, challenges, etc., and these approaches still have a place. But, as we continue to bring safety to new and higher levels within our organizations, we should consider balancing the equation, putting those we serve at ease, building a lasting relationship. We should consider handing a Safety Awareness Item (SAI) to another during a critical safety conversation…or when we are asking for a safety sale. If it worked for Joe to the count of 13,001 career new car sales and a Guinness World Record, it can help us sell safety, one employee at a time.

Handing Up Safety…There is an old saying, "In order to pull someone up, we have to first be on higher ground." When it comes to building safety awareness within our work force, we are attempting to bring them to a new level, to open their eyes, to obtain their undivided attention at least for a short time, so we can deliver a message that will be heard, understood and retained. Maybe the best, and most untapped method for achieving this end is the use of a SAI during these interactions. For the most efficient use of the SAI concept, keep these five principles in mind.

First, it's not about the item, it's about the relationship. Joe Girard didn't hand out $100 bills or new wristwatches. Instead, he used simple and practical items that served a purpose. The purpose was to balance the equation so that he could comfortably ask a critical question; will you spend this money on that car? With a SAI, it's not about the item. It's not an attempt to "buy" safety or purchase performance. It is simply an attempt to soften the perimeter of the person you are talking with so that you may discuss safety awareness in a manner that will be heard.

Keep items practical for peak interest. I have supported some work groups that purchased impractical or "silly" items. They were into sticky rubber frogs or mood rings. The practice of using SAIs in these locations quickly turned from effective to marginal. Of course, you will never please everyone but for best results with SAIs, consider staying close to these three-core groups: food, tools and more food. A practical SAI is one that most people can use and appreciate even if they already have one or two at home. For example, a small flashlight with the theme, "maintain a clear path." Or, a roll of duct tape can be used emphasizing the need to use the right tool for the right job. A simple *Crunch* candy bar may be handed to a skilled craft professional when reinforcing line of fire crunch situations. The use of practical SAIs will allow you to finalize that sale…and in the end, that is what it is all about.

Like any good thing, SAIs can be overused so formulate a plan within your location around what constitutes effective frequency. In most cases, that frequency will fall somewhere between an item a month and one item per quarter. When calculating an effective frequency of SAIs consider the exposures faced by your workforce, the effectiveness of current safety awareness programs, changes in the workforce such as summer peak, seasonal outages or maintenance schedules, at-risk act trend analysis and the introduction of new initiatives or safe work practices.

For SAIs to be effective, it's all about the handoff. In learning from the best, Joe Girard, we find that he handed an item to a client when they were in his office. What happens in the office? A sale! For a SAI to be effective it should occur when the safety coaching or feedback is taking place. Over time, I have worked with some groups that have coupled terrific SAIs with an insightful theme at a great point in time only to blow it through the handoff. Instead of personally handing the SAI to a co-worker immediately before the critical safety conversation, they simply pointed to the items in the box and told everyone to be sure to "pick one up." If we are taking the time, effort and energy to use the SAI concept, make sure we employ an effective handoff. I believe that immediately before asking for the sale (having that critical safety conversation) is the best time to hand out the SAI. Other effective times may include, greeting the workforce as they are entering the plant, waiting at the gate as employees are driving out, or in small group huddles as job planning is taking place.

Last, believe in the process. Years ago, Henry Ford coined a phrase, "Whether you believe you can or believe you can't, you are generally correct." The bottom line is, if you believe that a SAI will make a positive and impressionable contribution toward selling an individual on safety then you are correct. If you believe that this adds no additional value toward the current safety efforts, then you are also correct. The bottom line is that in sales, I mean safety, we need to leverage any minimal advantage we can. After all, the potential of failure (injury) is too permanent and life changing not to employ all possible measures. This process works. It shows caring on the part of the person giving the items and increases the openness to listen on the part of the person receiving the item. Believe in the process. And, if we believe we can make a positive difference through this concept...we will generally be correct!

Most car sales professionals would consider one sale a day a great average; Joe averaged six new car sales per day. In fact, Joe set a single day sales record of 18 automobiles. Joe was so popular he had to schedule 15-minute appointments for customers. His sales numbers continued to climb as his monthly retail mark was typically more than an entire dealership, 174 cars, and he holds the world record for most career new car sales, 13,001. If you ask Joe, he is quick to tell you that these were all individual new car sales, no wholesale and no fleet deals. In safety, it's the same deal. We sell our message one person and one appointment at a time. As safety professionals we must use the traditional sales approaches but we must take notes from some of the best too. That would include a highly successful sales professional with a desk draw full of items...in our business they are SAIs and well worth the effort.

Notes on Committees

Before beginning the checklist, it is probably a good idea that we are working from the same safety committee model. While we can all picture a power washer, there may be many different 'pictures' as to what a safety committee looks like. Below are notes on safety committees;

❑   For starters, this checklist is written with the idea that a safety committee needs to formally meet monthly. Informal meetings and sub committee meetings occur outside of this once-a-month meeting.

❑   A safety committee should be small enough to be effective yet large enough to represent the different parts of a work location. As a general rule, if you have more than 12 to 15 people on the committee, you may want to give serious consideration to breaking into two committees representing different parts of the operation.

❑   Length of term for each safety committee member should be defined. For example, each member will serve a two-year term with half of the terms expiring in one year, the other half the following year.

❑   Who should be on a committee? The criteria for a typical committee follows;

  •   Sponsor-a sponsor is a member of the senior leadership team. This individual is not involved in the 'nuts and bolts' of the committee but does attend a few meetings and is always available for questions, support and to advance the committee's agenda.

  •   Committee Facilitator-this is a line management position who is the 'glue' of the committee. This individual works behind the scenes to ensure that each committee person is involved and the work of the committee is getting completed in a timely fashion.

  •   Committee members representing the various parts of the organization. For example, if one plant has five different manufacturing processes all contributing to a final product, each one of the five process should be represented. The goal is to have each process, especially when there are unique hazards presented in each process, represented. Each of these committee members would then have a formal role in the committee as outlined below.

- Safety Staff members participate in each committee meeting as a support tool and subject matter expert. The facilitator, not the safety staff member, is the one responsible to ensure assignments are complete, roles filled, etc.

January-Taking the Committee out of the box!

(Note; this assumes a committee year as running January through December. If your committee is on a different cycle, please consider January as Month-one).

There is an old saying that reads, "how a team begins will often determine how that team will finish." If your team is to have success, they will need to start strong. As you read through the first month's activities and see 10 steps outlined here, you will realize that this is going to be a busy and challenging month. After all, the other work doesn't stop while the committee is meeting. That being said, planning for the new committee should begin several months in advance of this first meeting. And, resources and scheduling should be in place before this first meeting to ensure the proper time and resources are in place to support the committee. If not, much credibility will be lost early in the process and many times team effectiveness will suffer as a result. Start strong and finish well!

10 ITEMS TO ASSEMBLE FOR JANUARY:

STEP 1—ESTABLISH A SPECIFIC END STATEMENT FOR THE TEAM;

As noted above in the Tips for On Target Teams;

"If you don't know where you are going," Yogi Berra says, "you'll end up somewhere else!" We would never hail a cab and then tell the cab driver, "I'm not sure where I need to go, just go." Instead, when we get in a cab, we have a direction, purpose and an expected arrival time; after all, the meter is running. Unfortunately, too often we stick our safety teams in a cab and tell the driver 'just drive' because we fail to give them crystal clear direction. ***The best way to provide our teams with direction along the ride is through a specific end statement*** (Hackman). "This is a statement of team destination but not the route (means) in which to get there. Management's role is to give the team this CDC, the team's job is to use the provided tools, training, knowledge and talent to get there."

That being said, a specific end statement must be established for the team in advance of the first meeting. It should be vetted by the supporting management team and be meaningful in terms of direction both for the team and the organization. An example of a specific end statement is as follows;

"To eliminate the exposures and change the culture that leads to injuries, incidents, near miss events and at-risk acts."

STEP 2—AGENDA AND ROLE REVIEW PRE WORK;
Before the meeting begins, the management leader should develop an agenda and route the agenda to the committee members and management support team at least one week prior to the meeting.

The agenda should include the following items;
- Beginning and ending time of meeting
- Location
- General purpose of meeting
- Review of committee duties and roles of each team member
- Election of Chairperson
- Election of roles
- Establishment of team direction, with specific end statement included on agenda (reference Tips for on Target Teams article to learn about specific end statement).
- Establish Goals
- Name team

❑ Discuss Safety Committee MAP process and develop strategy and a plan to develop MAPs for the next year.

❑ Discuss team training and set a date for this training.

STEP 3—ROLES AND RESPONSIBILITIES; ELECTION OF POSITIONS;

The truth is that we are of the most value where we add the most value. Each committee member approaches the committee with strengths and talents. Find a role in which each of those talents are able to shine and you will find a highly functioning safety committee!

In the past, many committees have elected a chairperson and recorder with the remaining committee members just showing up for the donuts! The chairperson's role and recorder's responsibilities are very important, yet every person on the committee should have at least one official role. These roles are predetermined and assigned. While I suggest that a committee take some time to brainstorm potential roles and each member volunteer for at least one position, below is a list of safety committee roles;

A Chairperson is:

1. *A liaison between management and the Team.*
2. *Facilitator of Team meetings.*
3. *Inclusive, keeps every Team member involved by making smart work assignments to Team members.*
4. *Understanding of each Team member's strengths, weaknesses and the key attribute each member can bring to the Team.*
5. *Mindful of the details of the Team such as agendas, sub committees, MAPs, safety observations, etc., and makes sure that all of these activities are completed in a timely fashion.*
6. *Directs each meeting in the direction of the Team's goals and specific end statement.*
7. *A winner, making sure the entire Team is following work practices and striving for safety excellence.*
8. *Ensuring the effectiveness of the Team meetings by directing discussions to meet goals.*

The Management Representative is;

1. Present, as a quiet example of leadership and to ensure the team is being effectively supported from the management team.
2. A promoter of the Team through the management ranks.
3. An equal, not pulling 'rank' during a meeting.
4. Participating in each meeting and volunteering to serve on subcommittees.
5. Pushing the team to reach its potential.

General 'Code of Conduct' for safety committee members;

1. Think 'shadow of a leader' and strive to continually set a good example.
2. Always use and promote the use of required personal protective equipment.
3. Coach and give feedback in regards to safe and unsafe work practices or conditions and always be ready to take whatever necessary action is needed to 'fix' an unsafe act and/or condition.
4. Promote Engagement; constantly finding ways to keep supervisors and fellow workers involved in safety.
5. Support the team in active contribution.
6. Maintain a positive attitude and be determined to strive for safety excellence even when things are not going as I would like.

Other Potential Committee Member Roles:

Note; Each member of the Committee accepts a specific Roles;

- ❑ Recorder
- ❑ Safety Observation Coordinator
- ❑ Safety Meeting Specialist
- ❑ Safety Leadership Award and/or Safety Award Coordinator
- ❑ Reporter (to work group)
- ❑ Trend analysis specialist
- ❑ Job Briefing Coordinator
- ❑ MAP organizer
- ❑ Goal Tracker
- ❑ Book Keeper (budget)
- ❑ Open issues tracker/Problem of the month

STEP 4—ESTABLISH TEAM DIRECTION;

Discuss the specific end statement with the team. Allow for a brainstorming session (Brainstorming tips included in the next section, Step 5) on the support, training, goals, roles, etc. that will be needed to reach the end statement.

The team should brainstorm the types of activities that should be used to reach the specific end statement. Below is list of some more common activities. The team needs to consider these as well as their own list and pick out the best ingredients for success:

- ❑ SMART Goals (explained in next section)
- ❑ Monthly Awareness Plans (MAPs) as outlined in the MAP section of this book.
- ❑ Focus on Three —Identify the top three hazards, based on trend analysis, injury reports and safety observation data and build an awareness program to prevent those three.
- ❑ Purpose —Write a committee purpose statement
- ❑ End of Year Summary-Keep effective notes to track end of year accomplishments.
- ❑ Special Teams —Set aside sub committees, a sub committee on ergonomics for example, to drill down deeper on key issues.
- ❑ Safety observation team-Ensure the safety observations are not only completed each month, but also done well!
- ❑ Mission Statements-Explains how specific end statement will be accomplished.
- ❑ Culture Building Initiatives-What is the culture at your location and how could it be improved?
- ❑ Involved Safety Meetings-If people get up and do, they will retain and use…
- ❑ Personal Touches-A personal touch is a one on one safety meeting with a safety committee member and a employee in the work group; how many personal safety messages can your committee deliver each month?
- ❑ Safety Saves/Good Catch-Set a program to celebrate each time an at-risk act is corrected.

STEP 5—SET GOALS

We would never get in a cab and not tell the cab driver where we want to go. Goals set and establish the committee's direction.

Hints for Goal Setting;

Spend time brainstorming issues, problems, trends, data, personal interests and talents toward safety, etc. In terms of injury data, safety observation data and other important information, that will need to be prepared in advance and provided to the team.

After listing and 'white boarding' all ideas from the brainstorming session, ask the team to narrow the list to the three to seven of the most important ideas. Set goals around these items.

Set proactive goals. Goals should not be related to injury numbers or data. Goals should outline exactly what you plan to do. For example, a poor goal would be: *"Zero electrical contacts for the district."* That is a great idea, but without action items, the team still has little direction. A much better example follows:

The district committee will experience zero electrical contacts this year by:
1. Reviewing monthly incident summaries involving past contacts.
2. Each month, writing the recommendations from these incident summaries on a poster board and posting in at least six locations around the district.
3. Have refresher training on use of rubber gloves, personal protective grounds, equipment grounds and WPA.
4. Have monthly quizzes on safety rules pertaining to electrical safety.

This goal can be measured and tracked. At year's end, the committee will know for sure what was done to achieve the goal. Work to form your committee's goals in a similar fashion.

Establish three to five proactive goals. As one goal is completed, establish another.

Include your goals as the first discussion of each meeting, tracking progress and getting reports from those who had assignments regarding the goals.

Reference: Goals should be SMART;
- [ ] S - specific and significant
- [ ] M - meaningful and motivational
- [ ] A - agreed upon and action-oriented
- [ ] R - rewarding and results-oriented
- [ ] T - time based and trackable
(Source unknown)

Brainstorming Ground Rules
- [ ] Designate a set period of time (e.g., 10 minutes)
- [ ] Have participants generate an exhaustive list of causes using one of the following methods:
- [ ] Participants call out ideas while working as a group
- [ ] Participants take several minutes to write down their own ideas and then share with the group
- [ ] Avoid judging ideas while brainstorming. Don't discuss each idea in detail until *after* all ideas have been recorded
- [ ] When brainstorming, go for quantity! When finished, go back through the list and cross off those that seem completely unlikely
(Source unknown)

STEP 6—NAME YOUR TEAM:

This is an optional step but many teams have gained great individuality, identity and ownership by naming their team. Most of the time, the name carries a greater meaning. For example a team has been called the Green Lakes BEARS, BEARS stands for, Before Every Action Remember Safety. Or Regional WASP team, WASP means, Workers As Safety Personnel.

In addition to the name, team logos can be established and included on all agendas and printed materials.

STEP 7—KICK OFF THE YEAR!

As you know, serving on a safety committee is work! The purpose of the kickoff is for the entire committee to be educated, entertained, inspired and appreciated for the role of safety committee member. Along with the networks gained, you will receive tools and take-a-ways that can help make your committee the best it can be.

Hints on successful safety kickoffs;
- ❑ Hold the kickoff at the beginning of the year, when the new committee is starting.
- ❑ Form a planning committee for the kickoff.
- ❑ Establish a purpose and theme for the meeting—these will guide the planning committee to ensure they stay focused on the end goal.
- ❑ Allow senior management to have time during the kickoff to express both appreciation and expectations for the coming year.
- ❑ Look to outside speakers. This is a good opportunity to bring in at least one, if not two, outside speakers. Generally, you can find a good resource in the community who is a good fit given the purpose and theme. Think about a local coach, college athletic director, police chief, etc. Also, look to bring in a professional speaker. You can learn more about my speaking services at www. thesafetysoul.org
- ❑ Have tangible take-a-ways. These are tools that the committee can use for the coming year. The 'take-a-way' can be a book to give your committee ideas, such as *ISMA (Involved Safety Meeting Activates)—101 Ways to Get Your People Involved,* for example. In addition, hats, coats, gloves, etc. make very good give-a-ways too.

STEP 8—HOLD THE JANUARY MEETING

This book is written on the assumption that the safety committee will meet monthly. Sub committees are formed as needed for special projects and assignments and report back to the larger committee during these monthly meetings. To be as effective as possible, use a Safety Committee Meeting Template for each committee meeting. It's a guide to help ensure that all-important information is reviewed during committee meetings. An example of a Safety Committee Meeting Template is as follows:

Safety Committee Meeting Template:

Safety Committee Specific End Statement: _To eliminate the expo-sures and change the culture that leads to injuries, incidents, near miss events and at-risk acts…_

Administrative:
- Agenda-published five days in advance and routed to each commit-tee member
- Meeting Minutes-published five days after previous meeting and routed to committee and management.

Goals: begin each meeting with a goals review.
- Have goals printed on agenda to facilitate review.
- Review goals and status each goal based on set measures for achievement.
- Form an action plan on each goal for the next month.

Awareness Initiative:
- MAPs-Have a copy of the next month's MAP for each committee member
- Make assignments for MAP activities.
- Plan ahead, as appropriate, for the new MAPs
- Discuss and plan for other tools for awareness such as newsletters, awareness items, appreciation targets and activities, etc.

Subcommittee Reports:
- What subcommittees are active and ready to report?

Trend Analysis Report:
- Review Safety Observation Data
- Review work group 'pulse' as learned from Safety Observation.
- Review any monthly reports published by Safety Staff
- Review near misses and injury reports over the previous month
- Project a trend analysis and form a plan for next month as to awareness, tool improvement, training etc. to counteract identified trends.

Forward Thinking:
Brainstorm on one or all of the following; Improvement opportunities, Hazards to be eliminated, Awareness activities, Tool Improvements, etc.

Review Committee Old Business/New Business:

Meeting Summary:
- At the end of the meeting, include a wrap up to review actions items and all assignments.
- Give feedback. As the last exercise before adjourning, have each committee member suggest at least one constructive feedback and one appreciative feedback. To help in this activity, have the feedback framed around the following phrases;
    o I think the committee can be even more effective if...
    o What I appreciate about the committee is...

Set Next Meeting Time and Date:

Ask, is the committee focused on the 'real' issues, projects or goals that meet the specific end statement?

STEP 9—IDENTIFY AND RECEIVE COMMITTEE TRAINING.
Work with your respective safety supervisor, line manager and/or training professional to identify the training needed to ensure the safety committee's success. This training may include the following;
- Safety Observation training
- Incident Analysis
- Problem Solving
- Safety Committee MAP
- Involved Safety Meeting Activity/Effective Meeting Leadership

STEP 10—COMPLETE MEETING CRITIQUE FORM
It is no secret that feedback is the lifeblood of committee success. In other words, the committee can only grow, develop and ultimately yield results if given a steady diet of feedback. To that end, one key component

to committee success is a post meeting check sheet. This checklist should include key components of the meeting such as;

- ❑ Was there an agenda?
- ❑ Was the agenda followed?
- ❑ Were goals and the specific end statement reviewed?
- ❑ Was safety observation data discussed?
- ❑ Did each committee member arrive with assignments complete?
- ❑ Was a MAP completed?
- ❑ Is there a MAP prepared for the next month?
- ❑ Is the committee focused in the correct direction given any available injury/near miss data?

These are just a sampling of questions. The key is to develop a feedback sheet. After each meeting, have the management supervisor/Safety Staff participant or other neutral participant fill out the sheet and share it immediately after the meeting with either the committee chairperson or the entire committee. Adjustments should be made between the end of this meeting and the next meeting to improve in different areas.

EXTRA CREDIT!
—HINTS, TIPS AND TRICKS TO EVEN BETTER COMMITTEE RESULTS!

Budget—many safety committees are given a budget so that they are in control of certain expenditures like safety awareness items, printing of logos, professional fees for a outside safety speaker, etc.

I have seen this work two ways. First, the committee will be given a certain amount of money and told to 'make it work.' While it may not be the amount the committee may have asked for, it is a bucket of money and the committee is in control. The other approach I have seen is when a budget sub committee of the safety committee is formed. This budget committee brainstorms expenses and creates a line-itemed budget that is presented to the manager or director. If approved, this budget committee continues providing oversight for budget compliance.

February-3 Items to Assemble

STEP 1—PRE MEETING WORK

After the January meeting ends and before the February meeting, the management facilitator should ensure that the committee is moving forward. This would include;

- ☐ Publishing the meeting minutes from the January meeting no more than five days after said meeting.
- ☐ Sharing specific end statement, goals, MAP themes, roles, etc, with the work group the committee serves.
- ☐ Ensuring any and all subcommittees are supported. This would include time to meet and the resources needed to hold an effective meeting.
- ☐ Make sure all training is either complete or scheduled.
- ☐ Make sure an agenda for this meeting is prepared and routed, along with the meeting minutes, at least five days before the meeting.

STEP 2—HOLD MEETING

Use the meeting template printed in Step 8 of the January instructions to facilitate the meeting.

STEP 3—COMPLETE MEETING CRITIQUE FORM

Follow the instructions in step 10 of the January instructions and complete this critique form. Keep in mind that sharing with the safety committee chairperson is the key to improving the results of the committee.

EXTRA CREDIT!
—HINTS, TIPS AND TRICKS TO EVEN BETTER COMMITTEE RESULTS!

The management facilitator should evaluate the Safety Committee in terms of CCD (Crystal Clear Direction). CCD is broken down into Processes and Role Assignments.

Processes are the 'plays' that the committee will run in order to meet the specific end statement. Some of these processes can include; SMART Goals, MAP process, Good Catch concept, Safety Touches, Year end Award Task Team, Focus on Three, Purpose/Mission Statement, Issue of the Month, Involved Safety Meeting Activities, etc.

Role Assignment can include but is not limited to; Committee Chair, Recorder, Year-end Award Coordinator, Goal Tracker, MAP Coordinator, Open Issues Tracker, Safety Observation Director, Communications Director, Media Advisor etc.

Evaluate your committee's Crystal Clear Direction with key questions such as;

Does each member have a role?

What are the goals for the year?

How will the committee reach the Specific End Statement?

Give specific feedback to your respective Manager and Safety Committee Chair and coach your committee as needed based on the evaluation.

March-3 Items to Assemble

STEP 1—PRE MEETING WORK

After the February meeting ends and before the March meeting, the management facilitator should ensure that the committee is moving forward. This would include;

- ☐ Publishing the meeting minutes from the February meeting no more than five days after said meeting.
- ☐ Sharing specific end statement, goals, MAP themes, roles, etc, with the work group the committee serves.
- ☐ Ensuring any and all subcommittees are supported. This would include time to meet and the resources needed to hold an effective meeting.
- ☐ Make sure all training is either complete or scheduled.
- ☐ Make sure an agenda for this meeting is prepared and routed, along with the meeting minutes, at least five days before the meeting.

STEP 2—HOLD MEETING

Use the meeting template printed in Step 8 of the January instructions to facilitate the meeting.

STEP 3—COMPLETE MEETING CRITIQUE FORM

Follow the instructions in step 10 of the January instructions and complete this critique form. Keep in mind that sharing with the safety committee chairperson is the key to improving the results of the committee.

EXTRA CREDIT!
—HINTS, TIPS AND TRICKS TO EVEN BETTER COMMITTEE RESULTS!

Evaluate the Safety Committee you serve on the concept of Clear Structure and team Coaching. Remember, clear structure is hard to define but vital for team success. The best definition might simply be, "What's acceptable around here?" Evaluate 'what's acceptable' when it comes to effort and performance as well as the nuts and bolts of committees such as being driven by goals and specific end statements, agendas and meeting minutes published on time, members coming to meeting on time, prepared and focused, meetings starting on time, effectiveness of meetings etc. Give specific feedback to your respective Manager and Safety Committee Chair.

April-4 Items to Assemble

STEP 1—PRE MEETING WORK

After the March meeting ends and before the April meeting, the management facilitator should ensure that the committee is moving forward. This would include;

- ❑ Publishing the meeting minutes from the March meeting no more than five days after said meeting.
- ❑ Sharing specific end statement, goals, MAP themes, roles, etc, with the work group the committee serves.
- ❑ Ensuring any and all subcommittees are supported. This would include time to meet and the resources needed to hold an effective meeting.
- ❑ Make sure all training is either complete or scheduled.
- ❑ Make sure an agenda for this meeting is prepared and routed, along with the meeting minutes, at least five days before the meeting.

STEP 2—HOLD MEETING

Use the meeting template printed in Step 8 of the January instructions to facilitate the meeting.

STEP 3—COMPLETE MEETING CRITIQUE FORM

Follow the instructions in step 10 of the January instructions and complete this critique form. Keep in mind that sharing with the safety committee chairperson is the key to improving the results of the committee.

STEP 4—FORMALLY APPRECIATE THE COMMITTEE

If given the proper coaching, support, structure, training, etc. the committee is up and running and getting results! It's time to formally appreciate the committee. My favorite suggestion is to simply take the committee to lunch. If a member of senior management can join the team lunch, as an additional sign of support, that is positive as well.

EXTRA CREDIT!
—HINTS, TIPS AND TRICKS TO EVEN BETTER
COMMITTEE RESULTS!

Evaluate the Safety Committee on two concepts; Shared Belief System and Support.

"Understand," Max DePree, past CEO, Herman Miller Co. said, "What we <u>believe</u> precedes policy, procedure and practice." Are there beliefs holding your committee back from reaching its potential or the specific end statement? Evaluate and offer feedback.

Next, the five key areas of Safety Committee Support are: Time, Budget, Training/Skill Development, Clerical/Office Support and Feedback/ Recognition. Evaluate each of these support functions and decide if support is lacking in any of these areas.

Give specific feedback to your respective Manager and Safety Committee Chair.

May-4 Items to Assemble

STEP 1—PRE MEETING WORK

After the April meeting ends and before the May meeting, the management facilitator should ensure that the committee is moving forward. This would include;

- ☐ Publishing the meeting minutes from the April meeting no more than five days after said meeting.
- ☐ Sharing specific end statement, goals, MAP themes, roles, etc, with the work group the committee serves.
- ☐ Ensuring any and all subcommittees are supported. This would include time to meet and the resources needed to hold an effective meeting.
- ☐ Make sure all training is either complete or scheduled.
- ☐ Make sure an agenda for this meeting is prepared and routed, along with the meeting minutes, at least five days before the meeting.

STEP 2—HOLD MEETING

Use the meeting template printed in Step 8 of the January instructions to facilitate the meeting.

STEP 3—COMPLETE MEETING CRITIQUE FORM

Follow the instructions in step 10 of the January instructions and complete this critique form. Keep in mind that sharing with the safety committee chairperson is the key to improving the results of the committee.

STEP 4—PREPARE FOR MID-YEAR SAFETY SHARE

One great way to put wind in the sail of a safety committee and garner new and exciting ideas for your committee is to have a mid-year safety share. This would be a time when all of the safety committees from your company or organization come together and share. There are many forms an event like this can take but the general purpose is to have each group share what they have been doing, what seems to be working and what roadblocks they may have experienced. It is also beneficial for each committee to bring items, such as copies of effective safety meetings, MAPs, ISMA's and safety awareness items for others to have.

Work should begin by;
☐   Assigning a subcommittee to prepare for the event.
☐   Preparing a display table for the event.
☐   Acquiring safety materials that can be given away—this can include copies of MAPs, ISMA's and safety awareness items.

Note; if you do not have other safety committees in your company to share with, look outside your organization to other companies your same size and coordinate a 'city wide' safety share. A professional safety organization such as American Society of Safety Engineers (ASSE) may be of assistance. They probably have a local chapter that can provide support for such an event.

EXTRA CREDIT!
—HINTS, TIPS AND TRICKS TO EVEN BETTER COMMITTEE RESULTS!

Ask the Committee Chair to evaluate both the performance of the management facilitator and the overall support given to the team by the management team. Share the feedback with the management sponsor and safety staff member. Decide, based on the feedback, the steps needed to improve both the support from the facilitator and the management team.

June-4 Items to Assemble

STEP 1—PRE MEETING WORK

After the May meeting ends and before the June meeting, the management facilitator should ensure that the committee is moving forward. This would include;

- ☐ Publishing the meeting minutes from the May meeting no more than five days after said meeting.
- ☐ Sharing specific end statement, goals, MAP themes, roles, etc, with the work group the committee serves.
- ☐ Ensuring any and all subcommittees are supported. This would include time to meet and the resources needed to hold an effective meeting.
- ☐ Make sure all training is either complete or scheduled.
- ☐ Make sure an agenda for this meeting is prepared and routed, along with the meeting minutes, at least five days before the meeting.

STEP 2—HOLD MEETING

Use the meeting template printed in Step 8 of the January instructions to facilitate the meeting.

STEP 3—COMPLETE MEETING CRITIQUE FORM

Follow the instructions in step 10 of the January instructions and complete this critique form. Keep in mind that sharing with the safety committee chairperson is the key to improving the results of the committee.

STEP 4—CONTINUE TO PREPARE FOR MID YEAR SAFETY SHARE

The safety share is typically held in July. Have the subcommittee assigned to gather the material for the mid-year sharing show the committee, in this June meeting, what they plan to show. The committee can then offer feedback on the material.

EXTRA CREDIT!
—HINTS, TIPS AND TRICKS TO EVEN BETTER COMMITTEE RESULTS!

Gut-Check—assign a sub committee to summarize the first half of the year including the MAPs, goals, specific end statement, safety observation data trends, etc. Bring the summary to the July meeting as a means to benchmark committee performance in the first half of the year. This will also give the committee an opportunity to see clearly what they have ac-complished and set goals on the work that still needs to be done.

July-5 Items to Assemble

STEP 1—PRE MEETING WORK

After the June meeting ends and before the July meeting, the management facilitator should ensure that the committee is moving forward. This would include;

- [ ] Publishing the meeting minutes from the June meeting no more than five days after said meeting.
- [ ] Sharing specific end statement, goals, MAP themes, roles, etc, with the work group the committee serves.
- [ ] Ensuring any and all subcommittees are supported. This would include time to meet and the resources needed to hold an effective meeting.
- [ ] Make sure all training is either complete or scheduled.
- [ ] Make sure an agenda for this meeting is prepared and routed, along with the meeting minutes, at least five days before the meeting.

STEP 2—HOLD MEETING

Use the meeting template printed in Step 8 of the January instructions to facilitate the meeting.

STEP 3—COMPLETE MEETING CRITIQUE FORM

Follow the instructions in step 10 of the January instructions and complete this critique form. Keep in mind that sharing with the safety committee chairperson is the key to improving the results of the committee.

STEP 4—MID YEAR SAFETY SHARE DEBRIEFING

Either part of the committee, or the entire committee attends the midyear safety share. The purpose of this is to learn what other committees have done and to share what your committee has accomplished. In learning what the other committees have done, spend some time reviewing that list. After the review, pick at least three items that other committees have done that you and your committee should also do. These new activities can bring a spark to the second half of the year.

STEP 5—BEGIN PLANNING FOR THE NEW YEAR SAFETY KICKOFF

There is obviously a lot of work in planning a safety kickoff and in order to be successful, one has to start planning early. In this planning process, give serious consideration to;

- ☐ The purpose of the kickoff
- ☐ Location
- ☐ Give-a-ways, what tangible items will each attendee walk away with?
- ☐ Speakers, including local subject matter experts and safety keynote speakers

EXTRA CREDIT!
—HINTS, TIPS AND TRICKS TO EVEN BETTER COMMITTEE RESULTS!

Go Shopping. Safety Awareness items are important tools both for safety awareness and as conversation starters for conversations on safety compliance. Now that it is mid-year, it is sometimes hard to find ideas for awareness items. One of the best ways to get the juices flowing is to take the committee shopping. Carve out an hour and split the group into pairs. Have each pair take one store. The assignment is to walk the store and write down the name of any item under a certain dollar amount, three-dollars for example, that could make a great awareness item. Meet back after the shopping spree and compare notes. From this activity you will have a short list of effective awareness items.

August-5 Items to Assemble

STEP 1—PRE MEETING WORK

After the July meeting ends and before the August meeting, the management facilitator should ensure that the committee is moving forward. This would include;

- [ ] Publishing the meeting minutes from the July meeting no more than five days after said meeting.
- [ ] Sharing specific end statement, goals, MAP themes, roles, etc, with the work group the committee serves.
- [ ] Ensuring any and all subcommittees are supported. This would include time to meet and the resources needed to hold an effective meeting.
- [ ] Make sure all training is either complete or scheduled.
- [ ] Make sure an agenda for this meeting is prepared and routed, along with the meeting minutes, at least five days before the meeting.

STEP 2—HOLD MEETING

Use the meeting template printed in Step 8 of the January instructions to facilitate the meeting.

STEP 3—COMPLETE MEETING CRITIQUE FORM

Follow the instructions in step 10 of the January instructions and complete this critique form. Keep in mind that sharing with the safety committee chairperson is the key to improving the results of the committee.

STEP 4—FORMALLY APPRECIATE THE COMMITTEE

Yes, it's time to appreciate the committee, again! Another round of formal appreciation seems in order as the committee rolls into the 'home stretch' of the year. My favorite suggestion is to simply take the committee to lunch. If a member of senior management can join the team lunch, as an additional sign of support, that is positive as well.

STEP 5—BUILD A LIST OF POTENTIAL NEW MEMBERS

Begin identifying new members! Yes, it's hard to believe that it is already time to look for new members but it is better to start now and build the best team possible than fail to plan ahead. After identifying who is 'termed,' ask the safety committee who they think will be good members. Compare their feedback to your internal list.

EXTRA CREDIT!
—HINTS, TIPS AND TRICKS TO EVEN BETTER
COMMITTEE RESULTS!

Ask, what did we miss? By this time of year, the committee will seem well established and on target with their goals. It is at this very moment that there could be blind spots developing. In order to stay grounded take some time to ask key leaders and workers what the committee might be missing. You can also ask, what would you like us to do better or different. Look for the kernels of good ideas in all of the feedback.

September-4 Items to Assemble

STEP 1—PRE MEETING WORK

After the August meeting ends and before the September meeting, the management facilitator should ensure that the committee is moving forward. This would include;

- ☐ Publishing the meeting minutes from the August meeting no more than five days after said meeting.
- ☐ Sharing specific end statement, goals, MAP themes, roles, etc, with the work group the committee serves.
- ☐ Ensuring any and all subcommittees are supported. This would include time to meet and resources to hold an effective meeting.
- ☐ Make sure all training is either complete or scheduled.
- ☐ Make sure an agenda for this meeting is prepared and routed, along with the meeting minutes, at least five days before the meeting.

STEP 2—HOLD MEETING

Use the meeting template printed in Step 8 of the January instructions to facilitate the meeting.

STEP 3—COMPLETE MEETING CRITIQUE FORM

Follow the instructions in step 10 of the January instructions and complete this critique form. Keep in mind that sharing with the safety committee chairperson is the key to improving the results of the committee.

STEP 4—BEGIN FILLING OUT A FORMAL REVIEW SHEET ON THE COMMITTEE PERFORMANCE

If you don't already have an annual review form, make one! This is a key step to overall progress and results. The form should not only include a large section of safety committee performance, it should evaluate the entire work group that the committee serves. In the end, this is an effort to levelize committees across a set of criteria that are key to safety success. Some key evaluation items may include;

❑   How many injuries and incidences occurred?
❑   How many near miss reports filed? (In this case, the more the better).
❑   What was the number of vehicle crashes, if applicable?
❑   How can injury severity be measured?
❑   Is there an ergonomic program and what is the participation?
❑   Were there monthly Involved Safety Meeting Activities?
❑   Did the committee participate in the MAP process?
❑   Did the committee members have roles?
❑   Were committee minutes/agendas published on time and were they accurate?
❑   Did management and safety committee perform safety observations?
❑   Did the committee analyze this observation data?
❑   Did management give feedback to each worker regarding safety performance at least four times throughout the year?

This is just the beginning of the questions that can be used for an end-of-the year safety assessment. Again, the goal is to quantify performance over the past year.

EXTRA CREDIT!
—HINTS, TIPS AND TRICKS TO EVEN BETTER COMMITTEE RESULTS!

Begin making a survey? One way to understand what the work group would like to see from the committee and a way to include the work group is to survey or poll the group. Your committee may want to take this time to draft a survey. In the coming months, it can be shared with the committee and finalized. It is best to administer the survey in December taking the data from that tool to form goals in the New Year.

October-5 Items to Assemble

STEP 1—PRE MEETING WORK
After the September meeting ends and before the October meeting, the management facilitator should ensure that the committee is moving forward. This would include;

- ❑ Publishing the meeting minutes from the September meeting no more than five days after said meeting.
- ❑ Sharing specific end statement, goals, MAP themes, roles, etc, with the work group the committee serves.
- ❑ Ensuring any and all subcommittees are supported. This would include time to meet and resources to hold an effective meeting.
- ❑ Make sure all training is either complete or scheduled.
- ❑ Make sure an agenda for this meeting is prepared and routed, along with the meeting minutes, at least five days before the meeting.

STEP 2—HOLD MEETING
Use the meeting template printed in Step 8 of the January instructions to facilitate the meeting.

STEP 3—COMPLETE MEETING CRITIQUE FORM
Follow the instructions in step 10 of the January instructions and complete this critique form. Keep in mind that sharing with the safety committee chairperson is the key to improving the results of the committee.

STEP 4—FINALIZE NEW MEMBER LIST
Finalize the new member list and form a plan to ask for participation in the committee.

STEP 5—ESTABLISH SPECIFIC END STATEMENT FOR NEXT YEAR

In preparation for next year's committee, have the management team, along with consultation from the existing committee, write a specific end statement for the new committee, which officially begins in January.

EXTRA CREDIT!
—HINTS, TIPS AND TRICKS TO EVEN BETTER COMMITTEE RESULTS!

Collect all of the MAPs and ISMAs and put them in a book! This will be a great reference for the new committee and for line management, when responsible for safety meetings. This really works well if there are multiple committees. Then, a reference of ideas can be put together and used for the years to come!

November-6 Items to Assemble

STEP 1—PRE MEETING WORK

After the October meeting ends and before the November meeting, the management facilitator should ensure that the committee is moving forward. This would include;

- ☐ Publishing the meeting minutes from the September meeting no more than five days after said meeting.
- ☐ Sharing specific end statement, goals, MAP themes, roles, etc, with the work group the committee serves.
- ☐ Ensuring any and all subcommittees are supported. This would include time to meet and resources to hold an effective meeting.
- ☐ Make sure all training is either complete or scheduled.
- ☐ Make sure an agenda for this meeting is prepared and routed, along with the meeting minutes, at least five days before the meeting.

STEP 2—HOLD MEETING

Use the meeting template printed in Step 8 of the January instructions to facilitate the meeting.

STEP 3—COMPLETE MEETING CRITIQUE FORM

Follow the instructions in step 10 of the January instructions and complete this critique form. Keep in mind that sharing with the safety committee chairperson is the key to improving the results of the committee.

STEP 4—FINALIZE NEW MEMBER LIST

Send an invitation to new members so that they can attend the December meeting.

STEP 5—CONTINUE WORKING ON THE END OF THE YEAR ASSESSMENT.

Continue work on this assessment with the goal that it be turned in by year's end.

STEP 6—PURCHASE END OF YEAR APPRECIATION GIFTS FOR COMMITTEE MEMBERS.

It only seems logical that each committee member receive an appreciation gift for serving on the committee. Whether it is a shirt, hat, cooler, etc. offer them a token of your appreciation!

EXTRA CREDIT!
—HINTS, TIPS AND TRICKS TO EVEN BETTER COMMITTEE RESULTS!

Do something extra around the Holidays. Focus is such a large part of safety and the last two months of the year can pull at our 'on the job safety.' Consider an extra push, with the safety committee, to keep your work group focused on the hazards.

December-5 Items to Assemble

STEP 1—PRE MEETING WORK

After the November meeting ends and before the December meeting, the management facilitator should ensure that the committee is moving forward. This would include;

☐ Publishing the meeting minutes from the November meeting no more than five days after said meeting.

☐ Sharing specific end statement, goals, MAP themes, roles, etc, with the work group the committee serves.

☐ Ensuring any and all subcommittees are supported. This would include time to meet and resources to hold an effective meeting.

☐ Make sure all training is either complete or scheduled.

☐ Make sure an agenda for this meeting is prepared and routed, along with the meeting minutes, at least five days before the meeting.

STEP 2—HOLD MEETING

Use the meeting template printed in Step 8 of the January instructions to facilitate the meeting.

STEP 3—WELCOME NEW MEMBERS

The new members should be invited to this meeting, make them feel welcomed!

STEP 4—COMPLETE MEETING CRITIQUE FORM

Follow the instructions in step 10 of the January instructions and complete this critique form. Keep in mind that sharing with the safety committee chairperson is the key to improving the results of the committee.

STEP 5—TURN IN YEAR END ASSESSMENT

Share the final results of the assessment with the safety committee. This will give them great information and targets for their work in the coming year. After a full review of that report, turn it in to the appropriate person.

EXTRA CREDIT!
—HINTS, TIPS AND TRICKS TO EVEN BETTER COMMITTEE RESULTS!

Plan a celebration. If you have reached this point and checked off each box along the way, it has been a 'tough' yet rewarding year! It is now time to celebrate. Plan either a formal or informal celebration with the committee. Think about giving away some fun awards paying tribute to everyone's efforts over the last year.

Final Thoughts:

First of all, thanks for reading this book! Since 1997, I have been involved in many different sides of safety committees. In the late 1990s, I was a journey electrical line worker and served on a safety committee in Boonville, Missouri. A couple years later, I was a supervisor and a management facilitator for a committee. After that I was a member of the safety staff, supporting safety committees across Missouri in an advisory and motivational role. For the last several years, I have had the opportunity to consult with committees across the country, sharing with them and learning from them.

I mention this only because I know this book is just the start. I'd like the next book on safety committees to have not just my ideas, but your ideas too. I'd like to grow the monthly list of options, ideas and extra credit work. So as you are finishing this book, take a moment to add your experiences and ideas to each month and share them with me. I will collect these from safety professionals and leaders from across the country and share them in a new book. In so doing our committees and our safety community will grow exponentially.

Next, tell a friend. Seth Godin, in his insightful book entitled *Tribes*, responded to a question about how to sell books with the following two words, "Sell one." What he went on to explain is that if you can sell one book, that reader will tell ten people about it. Those ten people will read it and tell ten more people…and so on until the book is a huge success. "Or," he notes, "if the reader doesn't tell anyone then it's time to write a new book!" So, please tell a friend…tell ten friends! The point isn't to sell more books, although that is a nice upside. The point is to grow our safety community, share ideas and keep people safe. In the end, my company's mission and passion is to build people. Organizations with strong safety committees do just that; build people. And it's in this process of building safety committees and people where one will find great excitement, great rewards and great results!

Please share your thoughts, ideas and insights at www.thesafetysoul.org. Thanks again for reading this book. Matt.

Concepts take from the following Sources:
SawStop, LLC, 9564 S.W. Tualatin Road, Tualatin, OR 97062,
www.sawstop.com

Hackman, Richard J. *Leading Teams; Setting the Stage for Great Performances*,
Harvard Business School Publishing Corporation, 2002.

Meet Matt...

Matt Forck, CSP & JLW, is one of only a handful of individuals to earn both the Certified Safety Professional (CSP) designation and journey status in the hazardous field of distribution electrical line work. In addition, Matt has served as a manager in a corporate leadership team; the top 2.5% of a 3,800-member organization. It is through Matt's unique and diverse work experience that he has become a leading voice in safety, engaging organizations; from those that get their hands dirty, to safety professionals and top-level leadership.

Recently, Matt was dubbed 'The Safety Ambassador' due to his ability and passion to deliver proven, and results oriented keynote presentations throughout the United States. His style, which is life-light and laughter centered, is best summed up from an attendee at a recent conference, "Very energetic and fun! Everyone should attend his seminar, the best of the conference!"

Matt directs K-Crof Industries, LLC, an organization dedicated to "Building Safe People through writing, speaking and giving!" A sought after speaker and consultant, he appears at industry events and his corporate client list of includes fortune 500 companies, trade associations, utility and construction companies. Matt has published dozens of articles in various magazines and trade publications and authored six books. Materials for Safety Professionals including free resources and downloads may be found on the web; www.thesafetysoul.org.

Matt lives in Columbia, MO with his wife Stephanie, daughter Natalie, 8, and son Nathan, 7.

"Each year we look across the country and hand select the top 5% of speakers for our conference. Matt is one of those speakers!"
--Rick Donovan, Safety Professional. MO Mining Conference Planning Committee

K-Crof Products; Order Sheet:

Motivational and Inspirational Books;

❑ ***Check Up From the Neck Up-101 Ways to Get Your Head in the Game of Life.*** This book not only provides blueberries for the brain, it will help up get up and take action when life has us down.

❑ ***GUSTY -***The official "I can" handbook. This book gets you and your organization "in" to success.

Safety Related Books;

❑ ***Tailgate-101 Proven Stories to Begin Each Job Strong and Finish Safe!*** Most studies show that more than one-in-three injuries are due, in part, to poor job planning. This book provides motivational stories to launch any job briefing or safety meeting in the right direction.

❑ ***ISMA (Involved Safety Meeting Activities) –101 Ways to Get Your People Involved! -***It doesn't matter if you manage a safety sensitive job or a sales team, this book can change your training and meeting culture…

❑ ***The Safety Soul, Quotes and Stories to Inspire Safety and Life*** –What comes from the heart goes to the heart; and in safety we don't manage assets like tools or equipment, we manage relationships…this book can help make the difference.

❑ ***The Untapped Secret to Selling Safety -And 401½ Tangible Items Guaranteed To Help Make that Sale!*** Safety is about selling and this book will introduce one sales secret largely untapped and perfect for ongoing safety success.

Books may be purchased via the Internet: www.thesafetysoul.org or by calling K-Crof's offices, (573) 999-7981. Discounts for volume orders.

Presentations and Keynote Address; Contact Matt to learn more about his speaking and consulting services. Contact Matt directly via his website, www.thesafetysoul.org, email; matt@thesafetysoul.org or phone; 573.999.7981.

www.ingramcontent.com/pod-product-compliance
Lightning Source LLC
Chambersburg PA
CBHW082111070326
40689CB00052B/4595